Natural Resources

PLANTS

Lisa Bullard

DiscoverRoo
An Imprint of Pop!
popbooksonline.com

abdobooks.com

Published by Pop!, a division of ABDO, PO Box 398166, Minneapolis, Minnesota 55439.

Printed in the United States of America, North Mankato, Minnesota.

102019
012020

THIS BOOK CONTAINS RECYCLED MATERIALS

Cover Photo: iStockphoto
Interior Photos: iStockphoto, 1, 5, 6, 8, 11, 12, 13, 14, 17, 19, 21, 22 (top), 22 (bottom), 23 (bottom), 25, 26, 27, 29, 30, 31; Shutterstock Images, 7, 9, 15, 18, 23 (top), 28

Editor: Sophie Geister-Jones
Series Designer: Jake Slavik

Library of Congress Control Number: 2019942632

Publisher's Cataloging-in-Publication Data

Names: Bullard, Lisa, author.

Title: Plants / by Lisa Bullard

Description: Minneapolis, Minnesota : Pop!, 2020 | Series: Natural resources | Includes online resources and index.

Identifiers: ISBN 9781532165863 (lib. bdg.) | ISBN 9781532167188 (ebook)

Subjects: LCSH: Plants--Juvenile literature. | Natural resources--Juvenile literature. | Environment--Juvenile literature. | Ecology--Juvenile literature.

Classification: DDC 577.16--dc23

Pop open this book and you'll find QR codes loaded with information, so you can learn even more!

Scan this code* and others like it while you read, or visit the website below to make this book pop!

popbooksonline.com/plants

*Scanning QR codes requires a web-enabled smart device with a QR code reader app and a camera.

TABLE OF CONTENTS

CHAPTER 1

A WORLD OF AMAZING PLANTS

Tall trees form a thick roof over the rain forest. The sun barely shines through. Plants of many shapes spread underneath. Red flowers glow in green shadows. Ferns fan out over fallen leaves.

WATCH A VIDEO HERE!

Rain forests have four layers of growth. Specific plants grow in each layer.

Their smell fills the air. Insects fly from flower to flower. Rain forests are home to more life-forms than anywhere else on Earth.

There are nearly 400,000 kinds of plants in the world. They come in many shapes and sizes. Plants grow almost everywhere on Earth. They even grow on mountains and in Antarctica.

Arctic willow grows in very cold places.

Cacti are plants that live in hot, dry environments.

Flowers come in many shapes, colors, and sizes.

Many plants have roots, stems, and leaves. They produce flowers, fruits, and seeds to make new plants. Other plants, such as mosses, have different parts. But all plants are living things. Life on Earth depends on them.

Seagrass plants grow in large underwater groups. Some groups are so big that they can be seen from space.

CHAPTER 2

PLANTS ARE LIFE-GIVERS

Plants are one of Earth's most important resources. Plants supply food for people and animals. Food from plants includes fruits, nuts, and vegetables.

LEARN MORE HERE!

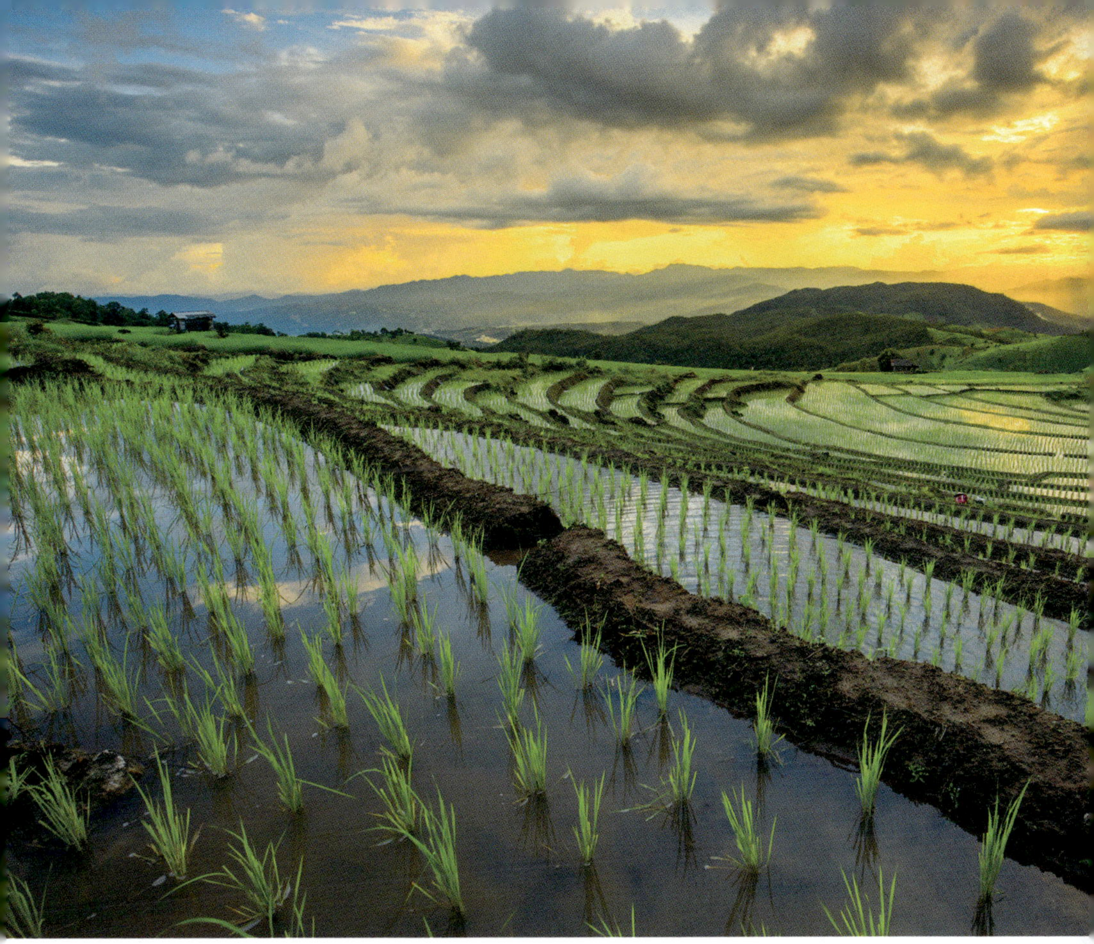

Rice plants grow in flooded fields. People have been eating rice for thousands of years.

A giant panda can eat 40 pounds (18 kg) of bamboo in a day.

Plants also help provide the world with **oxygen**. Unlike animals, plants make their own food. They use a process called **photosynthesis**. Plants take in sun, water, and gases in the air. They use these things to make energy. Then they

Rubber bands are made from trees.

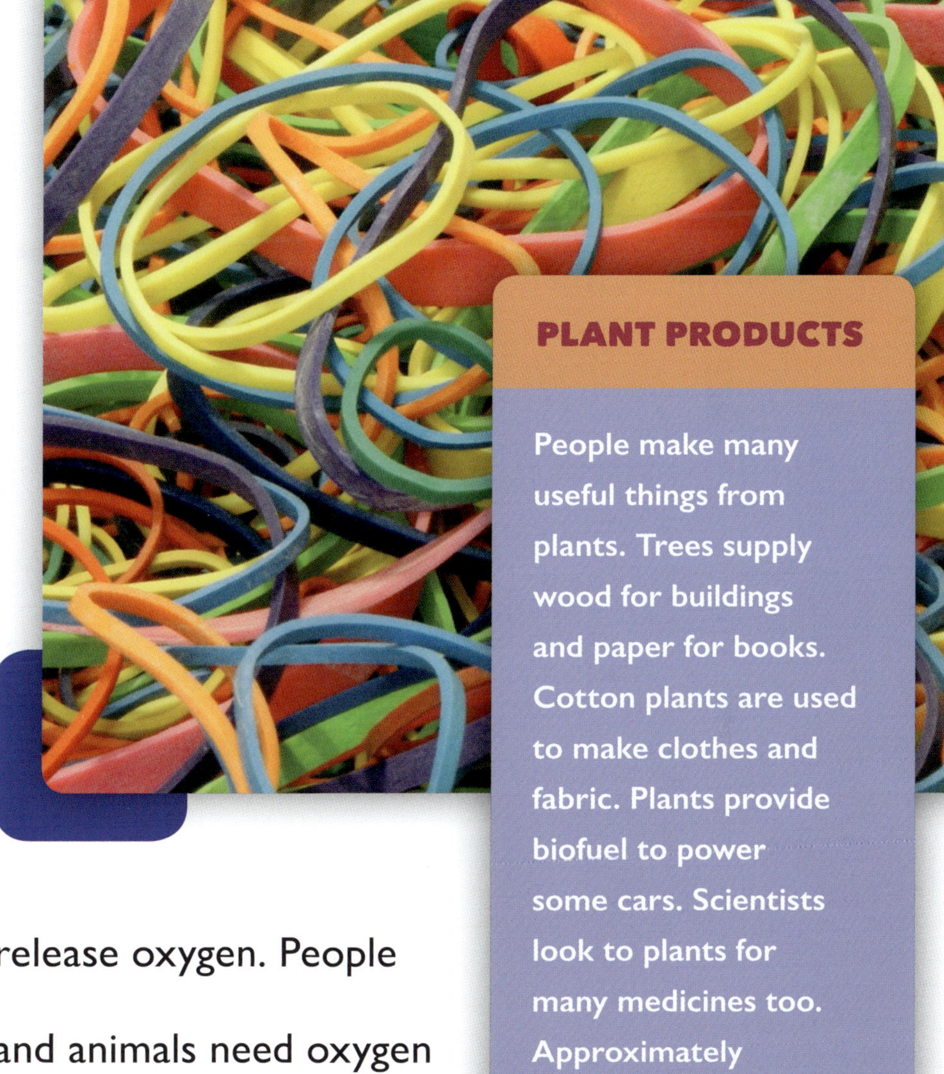

PLANT PRODUCTS

People make many useful things from plants. Trees supply wood for buildings and paper for books. Cotton plants are used to make clothes and fabric. Plants provide biofuel to power some cars. Scientists look to plants for many medicines too. Approximately 40 percent of medicines are plant-based.

release oxygen. People and animals need oxygen to live.

In addition, plants move water from the ground to the air. This process is called transpiration. A plant's roots draw water from the soil. Water vapor rises into the air from the leaves. This process is an important part of creating clouds.

People can place bags on plant leaves to study transpiration.

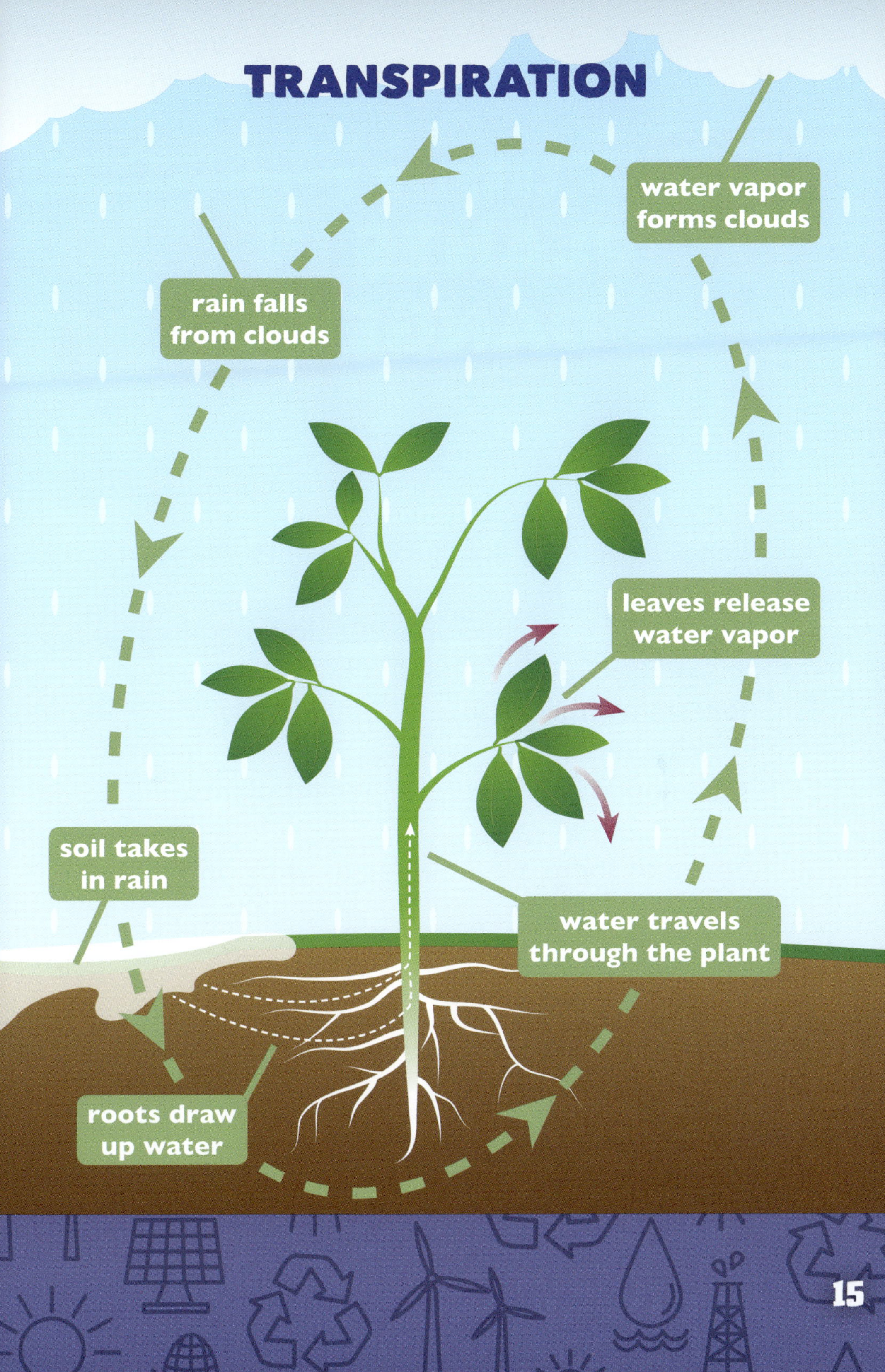
TRANSPIRATION
water vapor forms clouds
rain falls from clouds
leaves release water vapor
soil takes in rain
water travels through the plant
roots draw up water

CHAPTER 3

PLANTS IN DANGER

People's actions are hurting plants. As cities expand, people cut down forests to build houses or farms. Many areas once had several different plants. Some of them are now farms with only

COMPLETE AN ACTIVITY HERE!

New York City used to be a large oak forest.

one crop. This can create problems for animals. Bees, for example, need plants that bloom at different times.

Sometimes people use too much of certain plants. For example, people might cut down a whole forest to build houses from the wood. The animals who lived in the trees no longer have a home. Some kinds of animals might go **extinct**. Many trees grow slowly. It can take a long time for a forest to recover.

Most forests in Brazil where the Spix's macaw lived have been cut down.

In a method called clearcutting, all trees in an area are chopped down.

People sometimes grow plants from other environments in their gardens. However, these plants can become **invasive**. They can spread to places they do not belong. They often crowd out the area's **native** plants. The invasive plants take up all the sunlight, water, and growing space.

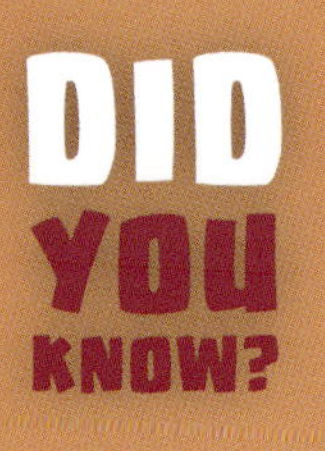

Scientists believe that at least one out of every five kinds of plant is in danger of going extinct.

Kudzu is an invasive plant that can grow 60 feet (18 m) in one year.

FROM ACORN TO OAK TREE

YEAR 1 (SPRING)

An acorn begins growing on a bur oak tree.

YEAR 1 (FALL)

The acorn falls to the ground and starts growing into a new tree.

YEAR 2

The tree is 1 foot (0.3 m) tall.

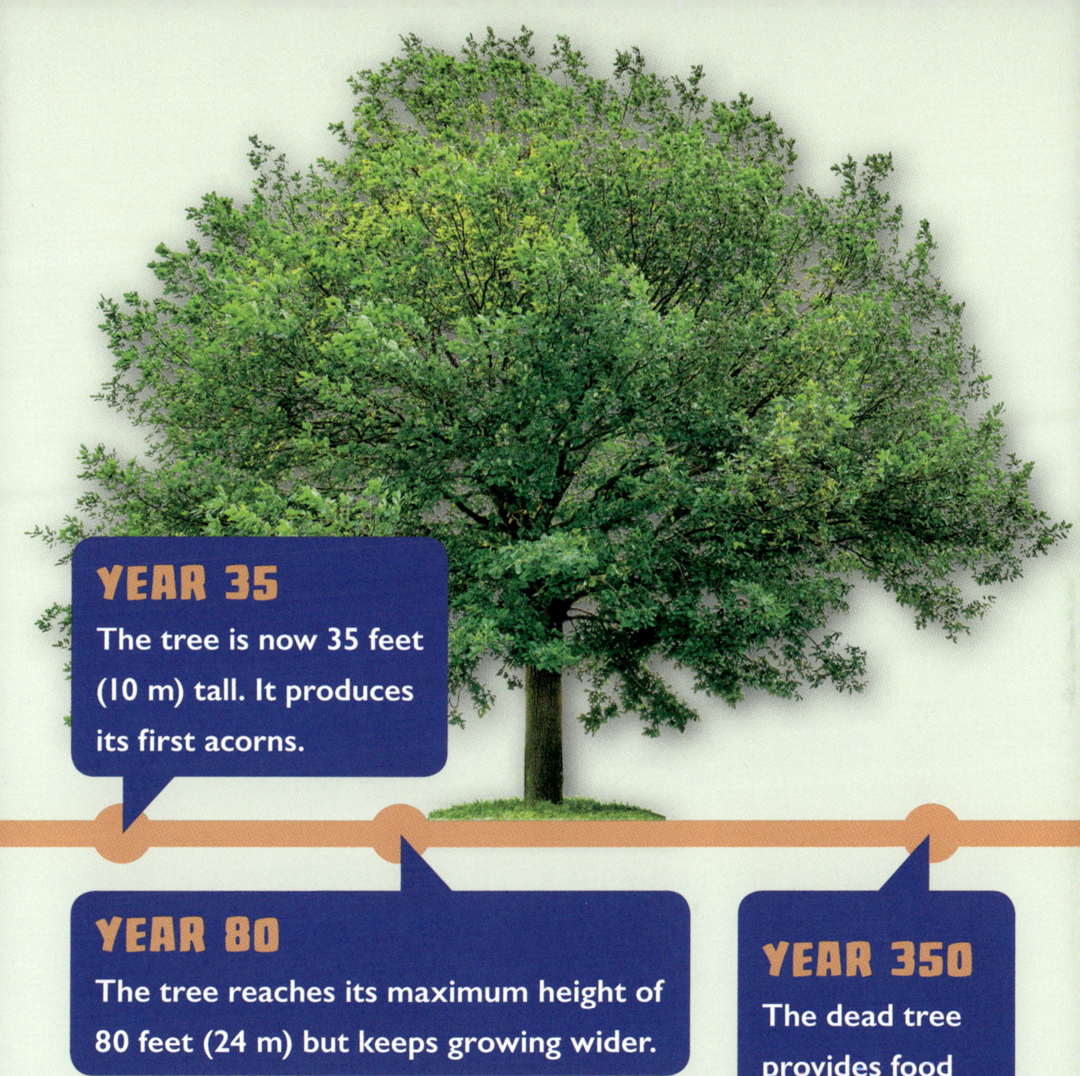

YEAR 35

The tree is now 35 feet (10 m) tall. It produces its first acorns.

YEAR 80

The tree reaches its maximum height of 80 feet (24 m) but keeps growing wider.

YEAR 350

The dead tree provides food and shelter for animals.

CHAPTER 4

KEEPING PLANTS SAFE

People around the world are working to help plants. Scientists study which plants are struggling. They study how to stop **invasive** plants and animals. Some scientists work with farmers. They look

LEARN MORE HERE!

Scientists who study plants are called botanists.

for ways to grow crops without hurting the environment. For example, planting different seeds can help.

Botanic gardens teach people about plants. They store plant seeds in seed banks. If one type of plant dies in the garden, scientists can replant its seeds. That way, it will not go **extinct**.

The Svalbard Global Seed Vault is the largest seed bank in the world.

Sequoia and Kings Canyon National Parks in California are known for their giant sequoia trees.

Governments are also working to help plants. They pass laws controlling the movement of plants between countries. These laws stop invasive plants. Governments also set aside areas, such as national parks, where plants can grow safely.

Everyone can help protect plants. Families can **recycle** paper. That means fewer trees are cut down.

In 2018, the United States recycled nearly 70 percent of the paper used.

People can also plant trees to replace the ones that were cut down. Other organizations are working to introduce **native** plants back to their natural environments.

Scientists have grown plants from very old seeds. Some are more than 1,000 years old.

MAKING CONNECTIONS

TEXT-TO-SELF

Have you ever planted something? What kind of plant would you like to grow?

TEXT-TO-TEXT

Have you read other books about plants? What kinds of plants did you learn about?

TEXT-TO-WORLD

Trees are cut down to produce paper. What is one way you or your family can use less paper?

GLOSSARY

botanic – relating to the scientific study of plants.

extinct – no longer existing.

invasive – coming from another place and causing harm.

native – naturally living in a certain area.

oxygen – a gas that people and animals need to breathe.

photosynthesis – a process where plants use light, carbon dioxide, and water to produce their own food.

recycle – to make trash into something that can be used.

INDEX

ONLINE RESOURCES

popbooksonline.com

Scan this code* and others like it while you read, or visit the website below to make this book pop!

popbooksonline.com/plants

*Scanning QR codes requires a web-enabled smart device with a QR code reader app and a camera.